ANIMALS IN DANGER

ELEPHANTS

Words that appear in BOLD are explained in the glossary.

First published by Hungry Tomato Ltd in 2022
F1, Old Bakery Studios, Blewetts wharf,
Malpas Road, Truro, Cornwall, Tr1 1QH, UK.

www.hungrytomato.com

ISBN 978 1 914087 83 7

Second edition

A CIP catalogue record for this book is
available from the British Library.

Printed in China.

Picture credits (t=top, b=bottom, m=middle,
l-left, r=right)

Superstock: 11br, 16-17bg. **Corbis**: 20-21bg.
Shutterstock: 10-11bg, 15br, 19br, 24br, 24tr,
25br, 25tm, 25tm, 32b, 4-5bg, 6-7bg, 7br; Anil
varma 26b; Bene_A 27bl; Claudio Soldi 28bl;
Efimova Anna 1bg 26tr; Erika J Mitchell 30br;
Flystock 22bl, 28bl; Four Oaks 30tr; Gregory
Zamell 22-23bg; Gregory Zamell 2bg, 8-9bg; Henk
Bogaard 12-13bg; Jez Bennett 27tr; Kikujungboy
CC 9br; Luison1976 3bg, 31bg; Marion Smith –
Byers 27mr; Michal Sanca 25tl; Mike Workman 28br;
Mikhail Kolesnikov 14-15bg; Mogens Trolle 28t; Paul
Hampton 27m; Pyty 24m;Sathienpong Prempetch 18-19bg;
Stockphoto mania 4bl; Tish1 12bl; Willyam Bradberry 25bl.

Every effort has been made to trace the copyright holders, and
we apologise in advance for any unintentional omissions. We
would be pleased to insert the appropriate acknowledgements in any
subsequent edition of this publication.

CONTENTS

THE BIGGEST OF THEM ALL

Elephants are the world's largest land-living animals.

These clever animals can survive in many different **habitats**, from the open **savannahs** of Africa, to the forests of Asia. Elephants can even survive in deserts.

Desert elephants in Namibia, Africa, may drink only once in three or four days. They are experts at searching for food over wide areas.

Elephants have no natural **predators** or enemies, but life is becoming difficult for them.

Asian elephant

DID YOU KNOW?

There are two types of elephant: African and Asian elephants.

A family of African elephants.

African elephants are larger than Asian elephants. They live in central and southern Africa, on open plains and dry deserts. Some of them live in forests.

There are two types of African elephant: African bush elephants and African forest elephants.

Both are **endangered**, but forest elephants are at the highest risk of **extinction.**

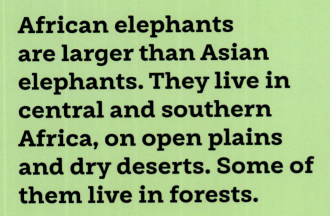

DID YOU KNOW?

Both male and female African elephants have tusks. They use them for digging up roots to eat.

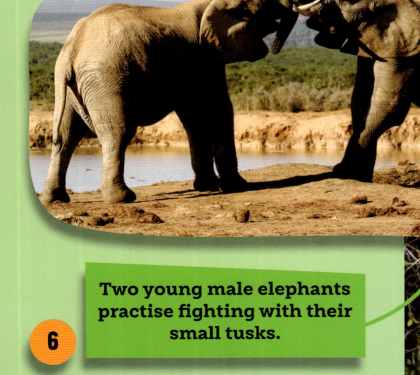

Two young male elephants practise fighting with their small tusks.

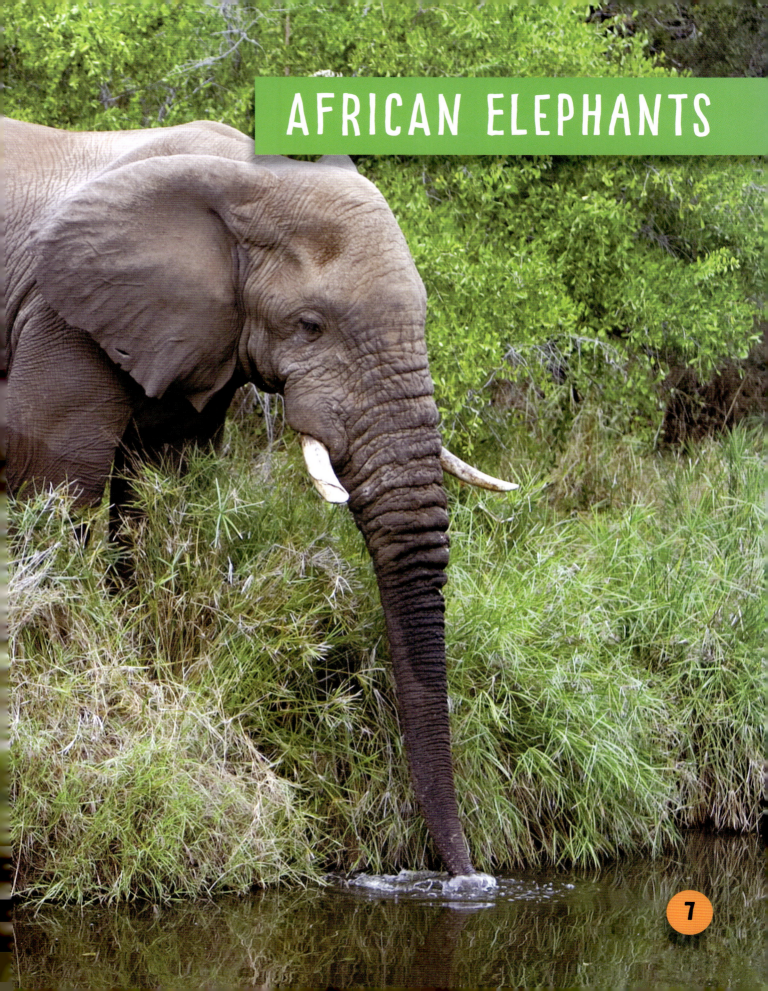

AFRICAN ELEPHANTS

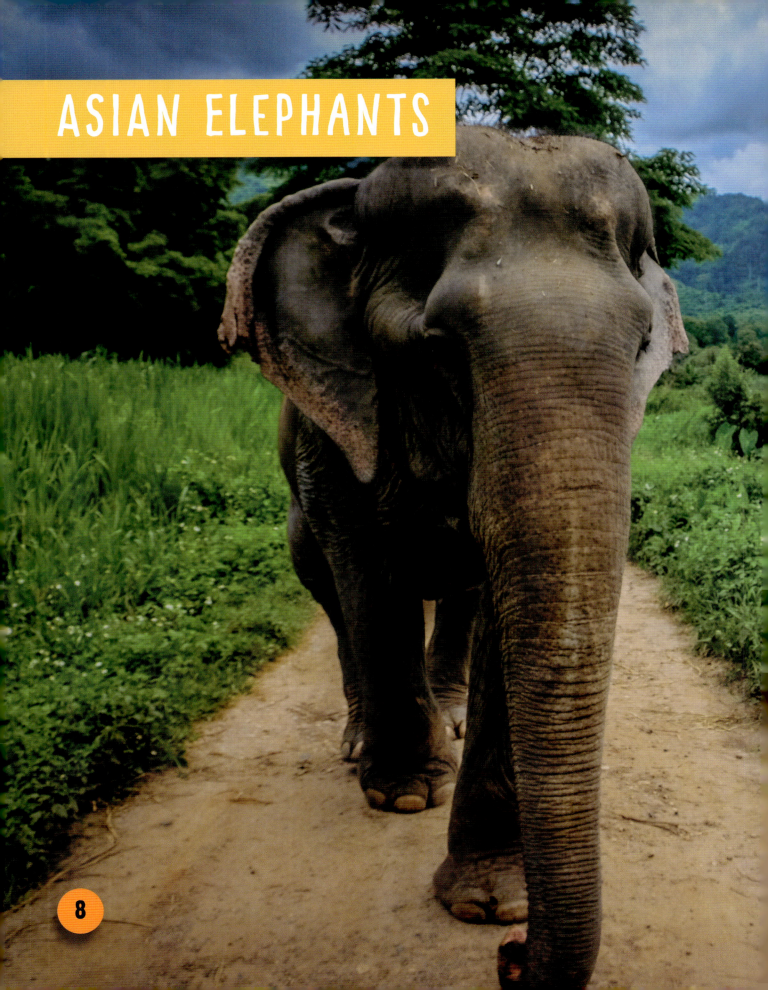

ASIAN ELEPHANTS

DID YOU KNOW?

Female Asian elephants don't have tusks.

Asian elephants live in many countries in South East Asia.

Most Asian elephants live in forests, but they are also found on **grasslands** and in mountain areas.

Asian elephants are smaller than African elephants. They have smaller ears and tusks, but the females do not grow tusks at all.

Asian elephants have been used as working animals for thousands of years. African elephants aren't traditionally used as working animals.

An Asian elephant being used as a working animal.

The calf in this picture has just been born!

Female and young elephants live together in family herds. All of the adults in the herd help to look after the babies.

The herds often join with others to make large groups called **clans.**

Males leave their herd when they are about 13 years old. They live alone, or in **bachelor groups** with other males.

Female elephants are called cows. They start to have calves when they are between 10 and 13 years old.

Elephants feed on different types of plants, including grass, leaves, branches, fruit, tree bark and even farm crops.

They eat a huge amount each day and need a big area to **forage** for food.

Each day an adult elephant drinks around 200 litres of water. Elephants are very clever at finding water. They can use their tusks to dig in the ground to find it.

Their digging makes waterholes that many other animals can use.

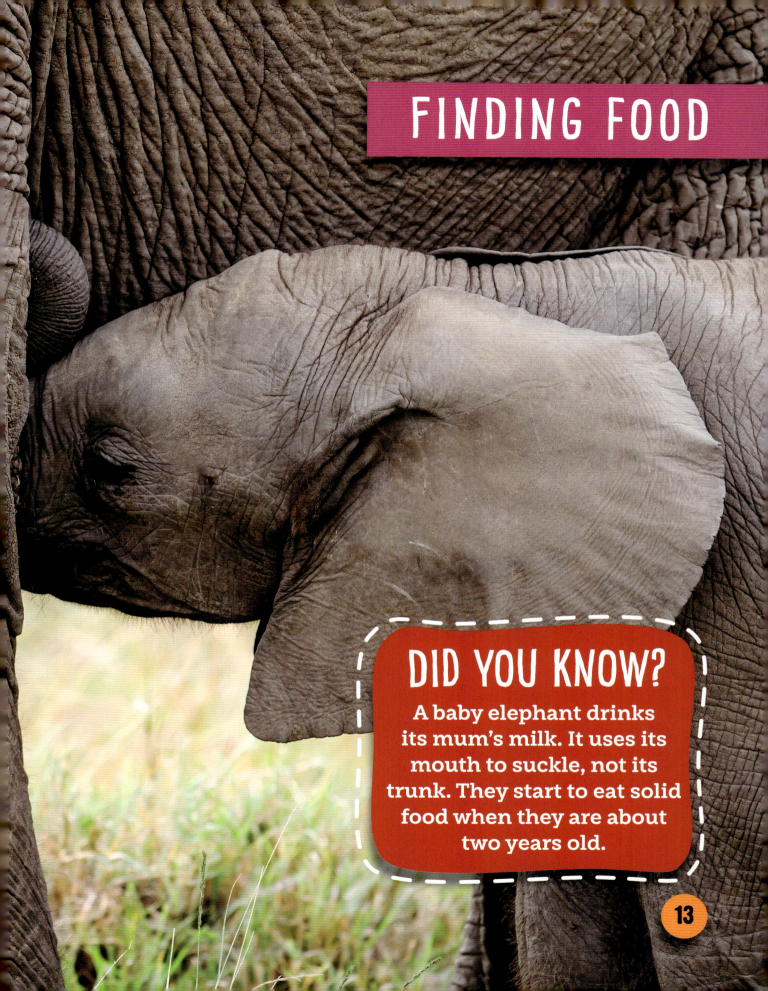

DID YOU KNOW?

A baby elephant drinks its mum's milk. It uses its mouth to suckle, not its trunk. They start to eat solid food when they are about two years old.

MAGNIFICENT TRUNKS

The 'fingers'

An elephant's trunk is an overgrown nose and lip.

The 'fingers' (lumpy bits) at the end of the trunk can tell whether an object is big or small, hot or cold, and what shape it is.

Trunks are very useful for lifting food into an elephant's mouth or for sucking up water to pour into its mouth.

Elephants often use their trunks like a shower to squirt water over their backs. Sometimes they blow dirt onto their backs for dust baths, to keep their skin healthy and cool.

An elephant having a dust bath.

TERRIFIC TUSKS

Tusks are very long teeth. If an elephant loses a tusk it will be difficult for it to dig for water, or roots to eat.

Baby elephants get their first real tusks when they are about one year old. Before that they grow tiny tusks, about 5 cm long. These are called 'milk tusks'.

Tusks are made of **ivory**. This is worth a lot of money.

DID YOU KNOW?

The largest elephant tusk ever recorded was 3.5 metres long and weighed 97 kg – the weight of a large man!

WORKING ELEPHANTS

Tourists getting rides on tame working elephants.

Elephants have been used as working animals for thousands of years. They are sometimes captured from the wild to be trained for this.

Asian elephants have been used to carry people and goods, lift heavy loads, pull carts and even carry soldiers into war.

Nowadays many working elephants are no longer needed because their work is done by machines.

In some countries, retired working elephants have to live in **sanctuaries** or **wildlife reserves** because they wouldn't know how to look after themselves in the wild.

DID YOU KNOW?

Some elephants have been trained to paint pictures! These are sold to raise money to look after retired working elephants.

HUNTED FOR IVORY

Game wardens have taken the tusks away from poachers. They then burn the tusks so they cannot be sold.

African elephants are hunted for food and for their ivory tusks.

Ivory is easy to carve, so in the past it was used to make beautiful objects that were worth a lot of money.

It is now against the law to buy or sell new objects made from ivory.

Nowadays elephants are protected and should not be hunted for ivory, but poachers ignore the laws and still kill elephants for their tusks.

DID YOU KNOW?

Just as humans are left or right handed, elephants are left or right tusked.

There is another danger to elephants. They are losing their habitat as it becomes farmland.

Forest elephants are threatened by **logging**.

Wildlife reserves can help give elephants a safe place to live. But they need huge areas to roam in, and lots of food.

If the reserves are too small, the elephants eat everything and there is not enough time for the plants to grow back.

Logging is cutting down trees for their wood.

A SAFE PLACE TO LIVE

WHERE DO ELEPHANTS LIVE?

ASIA

Asian elephants live in forests and open grassy places in the parts of Asia marked in red. Some live in mountain regions.

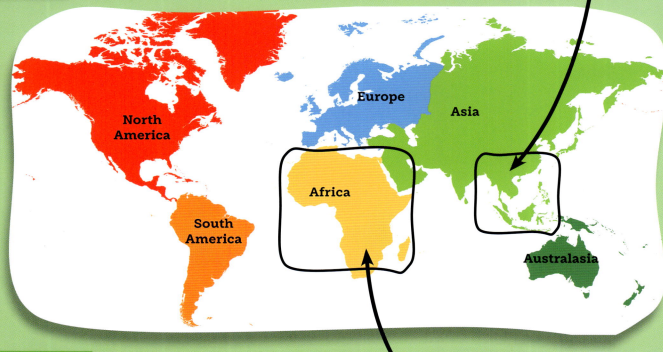

North America

Europe

Asia

Africa

South America

Australasia

AFRICA

African elephants used to live on savannahs and in forests across the whole of Africa. Now they only live in the area marked in red.

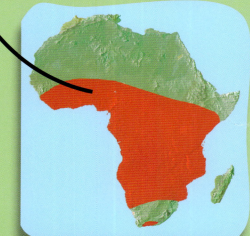

24

ELEPHANTS BODIES

Here are some amazing elephant facts:

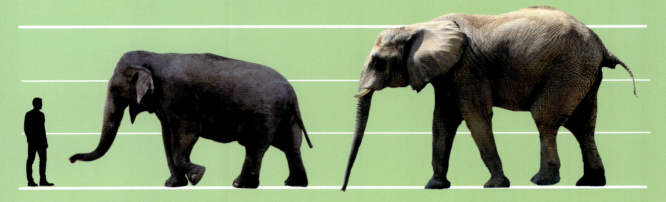

Asian elephant
Weight: 2,500 – 5,500 kg
Shoulder Height: 2.3 – 3 metres

African elephant
Weight: 1,800 – 6,600 kg
Shoulder Height: 2.5 – 4 metres

- Elephants can hold up to 6 litres of water in their trunk.

- Elephants can swim. They use their trunk like a snorkel!

- Elephants have wide, padded feet, which means they can walk very quietly.

ELEPHANTS IN DANGER

The two main risks to elephants today are poaching and habitat loss.

- The number of African forest elephants has decreased by a massive 86% in the last 31 years.

- In some areas, elephant numbers are starting to grow, but in others they continue to fall. There is still a lot of work to do to save them from extinction.

DID YOU KNOW?

There are fewer than 50,000 Asian elephants left in the world.

CONSERVATION

Wildlife conservationists are working hard to help save the elephants. Some of their work includes:

- Protecting elephant habitats by stopping people from being able to build and farm on it.

An elephant injured by poachers is helped by rangers.

- Researching and monitoring elephants to understand their behaviour and movement, and to keep track of their numbers and locations.

- Helping local people to live safely and peacefully alongside elephants.

- Protecting elephants from poachers and working to stop the illegal ivory trade.

HOW YOU CAN HELP

There are some great animal conservation charities and elephant sanctuaries that need money to pay for the work that they do. There are lots of fun ways that you could help to raise money and awareness for them.

- Why not organise a tasty cake sale?

- Ask people to sponsor you to do an activity. This could be something like a walk or a run, or helping to pick up litter in your local area.

- Put on a community or school event. How about a disco or a charity run?

Don't forget to get your family, friends and teachers involved! It'll be easier and much more fun with some help!

GLOSSARY

bachelor groups Groups of young male elephants.

clans A family group of elephants. The clan will include mothers, their babies, young males and females and the matriarch (the female leader).

conservation Taking care of the natural world.

conservationists People who work to protect animals and the natural world.

endangered At risk of extinction.

extinction When an animal or plant no longer exists anywhere on Earth.

forage To look for food in the wild.

game warden Someone whose job it is to look after wildlife reserves and the animals that live there.

grasslands Dry areas covered with grass where only a few bushes and trees grow.

habitats Places that suit particular animals or plants in the wild.

ivory The hard material that tusks are made from.

logging Cutting down trees for wood.

poaching The capturing or killing of animals by poachers so that they, or parts of their body, can be sold.

predators Animals that live by killing and eating other animals.

sanctuaries Safe places for animals that could not survive in the wild. If possible, the animals live a semi-wild life.

savannahs Large, open areas of land in Africa where grasses and bushes grow.

tourists People who are visiting on holiday.

wildlife reserves Places set aside for wild animals and plants to live. The animals and their habitat are protected by laws.

INDEX